ISBN 978-1-334-04269-0
PIBN 10721278

This book is a reproduction of an important historical work. Forgotten Books uses
state-of-the-art technology to digitally reconstruct the work, preserving the original format
whilst repairing imperfections present in the aged copy. In rare cases, an imperfection in
the original, such as a blemish or missing page, may be replicated in our edition. We do,
however, repair the vast majority of imperfections successfully; any imperfections that
remain are intentionally left to preserve the state of such historical works.

English
Français
Deutsche
Italiano
Español
Português

www.forgottenbooks.com

Mythology Photography **Fiction**
Fishing Christianity **Art** Cooking
Essays Buddhism Freemasonry
Medicine **Biology** Music **Ancient
Egypt** Evolution Carpentry Physics
Dance Geology **Mathematics** Fitness
Shakespeare **Folklore** Yoga Marketing
Confidence Immortality Biographies
Poetry **Psychology** Witchcraft
Electronics Chemistry History **Law**
Accounting **Philosophy** Anthropology
Alchemy Drama Quantum Mechanics
Atheism Sexual Health **Ancient History**
Entrepreneurship Languages Sport
Paleontology Needlework Islam
Metaphysics Investment Archaeology
Parenting Statistics Criminology
Motivational

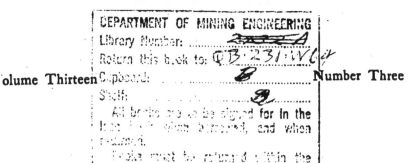
olume Thirteen   Number Three

QUARTERLY

OF THE

COLORADO
SCHOOL OF MINES

JULY, 1918

Issued Quarterly by the Colorado School of Mines
Golden, Colorado

Entered as Second-Class Mail Matter, July 10, 1906, at the Postoffice at
Golden, Colorado, under the Act of Congress of July 16, 1894.

Volume Thirteen

Number Three

QUARTERLY

OF THE

COLORADO ·
SCHOOL OF MINES

JULY, 1918

Issued Quarterly by the Colorado School of Mines
Golden, Colorado

Entered as Second-Class Mail Matter, July 10, 1906, at the Postoffice at
Golden, Colorado, under the Act of Congress of July 16, 1894.

QUARTERLY

OF THE

COLORADO SCHOOL OF MINES

| Vol. Thirteen | JULY, 1918 | Number Three |

Common Methods of Determining Latitude and Azimuth Useful to Engineers and Surveyors

COMPILED BY HARRY J. WOLF.

Professor of Mining.

I. LATITUDE.

1. By Observing Altitude of the Sun at Noon.

(a) Set up transit before local apparent noon. The standard time corresponding to local apparent noon at the point of observation may be found by adding or subtracting from 12 h the equation of time as directed in the Nautical Almanac or Solar Ephemeris.

(b) Find the maximum altitude of the upper or lower limb of the sun by keeping the middle horizontal cross hair tangent to the limb as long as it continues to rise. When the observed limb begins to drop below the cross hair read the vertical angle.

(c) Level the telescope and determine the index error. Apply this error to the observed vertical angle to obtain the true vertical angle.

(d) From a table of refractions in altitude determine the refraction correction for the vertical angle obtained, and subtract this correction from the true vertical angle to obtain the altitude of the limb observed.

(e) From a table of semi-diameters of the sun determine the semi-diameter for the date of observation, and add this correction if the lower limb was observed, or subtract it if the upper limb was observed, to obtain the altitude of the sun's center.

(f) From a table of the sun's parallax determine the parallax for the observed altitude, and add this correction to obtain the true altitude of the sun's center. In view of the limits of accuracy of the surveyor's transit this correction is usually neglected.

(g) From the Nautical Almanac or Solar Ephemeris determine the sun's declination at the instant the altitude was taken. If the longitude of the place is known, increase or decrease the declination for the instant of Greenwich apparent noon by the hourly change multiplied by the number of hours in longitude. If the longitude is not known, but standard time is known, increase or decrease the declination for the instant of Greenwich mean noon by the hourly change multiplied by the number of hours since Greenwich mean noon.

(h) Latitude $= 90° -$ Altitude $+$ N. Declination, or

Latitude $= 90° -$ Altitude $-$ S. Declination.

EXAMPLE:

(a) Transit was set up before local apparent noon on June 26, 1918. From Solar Ephemeris the equation of time is 2 m 27.73 s, and the difference for 1 h is 0.525 s. Transit was set up about 11:50 A. M., which was 12 m to 13 m before the sun's meridian transit.

(b) Upper limb of the sun was observed.

The vertical angle was =	73° 54′ 30″
(c) Telescope was leveled. Index error =	30″
True vertical angle =	73° 54′ 0″
(d) The refraction correction was =	17″
Altitude of sun's upper limb =	73° 53′ 43″
(e) Sun's semi-diameter was =	15′ 46″
Altitude of sun's center	73° 37′ 57″
(f) Sun's parallax was =	3″
True altitude of sun's center =	73° 38′ 0″

(g) Sun's apparent declination at Greenwich apparent noon on June 26, 1918 =   N. 23° 23′ 14.2″

The longitude of the place is 105° 32′ 33″, which = 7.04 h (15° = 1 h). Difference in declination = 7.04 x 4.38″ =   —30.8″

Sun's apparent declination at point of observation =   N. 23° 22′ 43.4″

(h) Compute latitude by formula: $Lat = 90° - Alt + N. Dec.$

	90″ 0′ 0″
Subtract altitude =	73° 38′ 0″
	16° 22′ 0″
Add N. Declination =	23° 22′ 43″
Latitude of place =	N. 39° 44′ 43″

2. By Observing Altitude of Polaris at Culmination.

(a) Set up transit before upper or lower culmination. The standard time of culmination may be found by interpolation from a table of time of culmination of Polaris in the Nautical Almanac.

(b) Focus on the star and follow it with the horizontal cross hair as long as it continues to rise if upper culmination is observed, or as long as it continues to fail if lower culmination is observed. When the desired culmination is reached read the vertical angle.

(c) Level the telescope and determine the index error. Apply this error to the observed vertical angle to obtain the true vertical angle.

(d) From a table of refractions in altitude determine the refraction correction for the vertical angle obtained, and subtract this correction from the true vertical angle to obtain the altitude of the star.

(e) From the Nautical Almanac or Ephemeris determine the polar distance of Polaris, either from a table of polar distances or by subtracting the apparent declination from 90°.

(f) Latitude = Altitude of the Pole.

Latitude = Altitude of Polaris at upper culmination — polar
       distance.

Latitude = Altitude of Polaris at lower culmination + polar
       distance.

EXAMPLE:

(a) Transit was set up before lower culmination on June 1, 1918. From a table of culminations of Polaris, the local mean time of lower culmination is 8 h 51.3 m P. M. for longitude 90° W. For longitude 105° 32′ 33″ W. the time would be 8 h 51.1 m P. M. (0.16 m earlier for each 15°). The transit was set up about 8:30 P. M. which is about 20 m be_fore lower culmination.

(b) Observed vertical angle =	38° 39′	0″
(c) Index error =	1′	0″
True vertical angle =	38° 38′	0″
(d) Subtract refraction correction =	1′	13″
Altitude of Polaris =	38° 36′	47″
(e) Polar distance on June 1, 1918	1° 8′	0″
Latitude of place (altitude of N. Pole). =	N. 39° 44′	47″

II. AZIMUTH.

1. By Observing Altitude of the Sun.

(a) Observe the sun at any time except when it is within 10° of the horizon (because the refraction is relatively large and uncertain) or when it is near the meridian (because small errors in observed altitude pro_duce relatively large errors in azimuth). Set up transit over one end of the line whose azimuth is desired. Sight along the line with the verniers set at 0°. With the lower clamp tightened and the upper clamp loosened sight on the sun with a colored shade glass on the eye piece or focus the sun's disc, and the cross hairs of the instrument, on a screen held behind the eye piece.

If the observation is made in the forenoon place the sun's disc in the upper left-hand quadrant, and tangent to the vertical and middle horizontal cross hairs, and record the vertical and horizontal angles and the time. Then reverse the instrument and make similar observations with the sun's disc in the lower right hand quadrant. If the observation is made in the afternoon, place the sun's disc first in the upper right-hand quadrant and then, with the instrument reversed, in the lower left-hand quadrant. The mean of the vertical angles and the mean of the horizontal angles may be assumed to correspond to the position of the sun's center at the instant indicated by the mean time reading.

The direct and reversed observations should be made within a short period of time, say 2 or 3 minutes. If the instrument is in perfect adjust_ment, the observation may be simplified by centering the intersection of the vertical and middle horizontal cross hairs on the sun's disc, with the assistance of diagonal cross hairs, stadia hairs, or concentric circles placed on a screen upon which the sun's disc is focused.

(b) From a table of refractions in altitude determine the refraction correction for the mean vertical angle of the sun's center, and subtract this correction from the vertical angle to obtain the altitude of the sun's center.

(c) From a table of the sun's parallax determine the parallax for the observed altitude, and add this correction to obtain the true altitude of the sun's center. In view of the limits of accuracy of the surveyor's transit this correction is usually neglected.

(d) From the Nautical Almanac or Solar Ephemeris determine the sun's declination at the instant the altitude was taken. If the longitude of the place is known, increase or decrease the declination for the instant of Greenwich apparent noon by the hourly change multiplied by the number of hours in longitude. If the longitude is not known, but standard time is known, increase or decrease the declination for the instant of Greenwich mean noon by the hourly change multiplied by the number ·of hours since Greenwich mean noon.

(e) The azimuth of the sun from the NORTH may be computed from any one of the following formulæ:

Where A_n = sun's azimuth from north

$$S = \tfrac{1}{2} \,(\text{codec} + \text{colat} + \text{coalt})$$

(1) $\quad \sin \tfrac{1}{2}\, A_n = \sqrt{\dfrac{\sin(S - \text{colat})\ \sin(S - \text{coalt})}{\sin \text{colat} \sin \text{coalt}}}$

(2) $\quad \cos \tfrac{1}{2}\, A_n = \sqrt{\dfrac{\sin S \ \sin(S - \text{codec})}{\sin \text{colat} \sin \text{coalt}}}$

(3) $\quad \tan \tfrac{1}{2}\, A_n = \sqrt{\dfrac{\sin(S - \text{colat})\ \sin(S - \text{coalt})}{\sin S \ \sin(S - \text{codec})}}$

Or from any one of the following formulae:

where A_n = sun's azimuth from north

$$s = \tfrac{1}{2} \,(\text{codec} + \text{lat} + \text{alt})$$

(4) $\quad \sin \tfrac{1}{2}\, A_n = \sqrt{\dfrac{\sin \tfrac{1}{2}\,(\text{lat} + \text{coalt} - \text{dec})\ \cos \tfrac{1}{2}\,(\text{lat} + \text{coalt} + \text{dec})}{\cos \text{lat} \sin \text{coalt}}}$

(5) $\quad \sin \tfrac{1}{2}\, A_n = \sqrt{\dfrac{\sin(s - \text{alt})\ \sin(s - \text{lat})}{\cos \text{lat} \cos \text{alt}}}$

(6) $\quad \cos \tfrac{1}{2}\, A_n = \sqrt{\dfrac{\cos s \ \cos(s - \text{codec})}{\cos \text{lat} \cos \text{alt}}}$

(7) $\quad \tan \tfrac{1}{2}\, A_n = \sqrt{\dfrac{\sin(s - \text{lat})\ \sin(s - \text{alt})}{\cos s \ \cos(s - \text{codec})}}$

(8) $\quad \text{vers } A_n = \dfrac{\cos(\text{lat} - \text{alt}) - \sin \text{dec}}{\cos \text{lat} \cos \text{alt}}$

The azimuth of the sun from the SOUTH may be computed from·any one of the following formulae:

where A_s = sun's azimuth from south

$$S = \tfrac{1}{2} \,(\text{codec} + \text{colat} + \text{coalt})$$

(9) $\quad \sin \tfrac{1}{2}\, A_s = \sqrt{\dfrac{\sin(S - \text{codec})\ \sin(S - \text{colat}}{\sin \text{codec} \sin \text{colat}}}$

(10) $\quad \cos \tfrac{1}{2}\, A_s = \sqrt{\dfrac{\sin S \ \sin(S - \text{coalt})}{\sin \text{codec} \sin \text{colat}}}$

(11) $\quad \tan \tfrac{1}{2}\, A_s = \sqrt{\dfrac{\sin(S - \text{codec})\ \sin(S - \text{colat})}{\sin S \ \sin(S - \text{coalt})}}$

(12) $\quad \cos A_s = \dfrac{\pm \sin \text{dec}}{\cos \text{lat} \cos \text{alt}} - \tan \text{lat} \tan \text{alt}$

Note—If the observation is made north of the equator the declination is + when north and — when south. If the observation is made south of the equator the declination is + when south and — when north. If the sun is observed when north of the prime vertical in the northern hem. isphere, or south of the prime vertical in the southern hemisphere, the first term will be greater than the second term. Equation (13) is an. other form of equation (12).

$$(13) \quad \cos A_s = \frac{\pm \sin \text{ dec} - \sin \text{ lat} \sin \text{ alt}}{\cos \text{ lat} \cos \text{ alt}}$$

$$(14) \quad \text{vers } A_s = \frac{\cos(\text{lat} + \text{alt}) + \sin \text{ dec}}{\cos \text{ lat} \cos \text{ alt}}$$

The azimuth of the sun from the NORTH may be computed from the following formulae:

where A_n = sun's azimuth from the north

$$(15) \quad \cos A_n = \tan C_1 \tan \text{ lat} = \tan C_2 \tan \text{ lat}$$

$C_1 = \frac{1}{2} \text{ coalt} + \frac{1}{2}(C_1 - C_2)$

when latitude is less than declination and on the same side of the equator.

$C_2 = \frac{1}{2} \text{ coalt} - \frac{1}{2}(C_1 - C_2)$

when latitude is greater than declination and on the same side of the equator, or when latitude and declination are on opposite sides of the equator.

$\tan \frac{1}{2}(C_1 - C_2) = \cot \frac{1}{2}(\text{lat} + \text{dec}) \tan \frac{1}{2}(\text{lat} - \text{dec}) \cot \frac{1}{2} \text{ coalt}$

EXAMPLE:

(a) Transit is set up over B.M. on June 4, 1918.
   Sighted on Flagstaff with verniers at 0°.
   Telescope pointed at sun, and the following observations recorded

Quadrant	Time.	Horizontal Angle	Vertical Angle
Upper right........	2:52 P. M.	294° 15′	50° 3′
Lower left.........	2:54 P. M.	295° 34′	49° 6′
Sun's center........	2:53 P. M.	294° 54′ 30″	49° 34′ 30″

(b) Refraction correction =   49″

Altitude of sun's center =   49° 33′ 41″

(c) Parallax correction =   6″

True altitude of sun's center =   49° 33′ 47″

(d) 1. If the Solar Ephemeris gives the sun's declination at Green-wich MEAN noon proceed as follows:

Sun's apparent declination at Greenwich MEAN noon on June 4, 1918 = N. 22° 22′ 22.0″. The difference in declination for 1 h = 18.03″.

The place of observation is west of longitude 105° W. and 105th meridian time is used. 105° = 7 h. The standard time of the observation was 2 h 53 m P. M. = 2.883 h, which is 7 h + 2.883 h = 9.883 h after Greenwich Mean Noon. The difference in declination at the instant of observation is 18.03″ x 9.883 = 178.2″ = 2′ 58.2″. The declination at the instant of observation is 22° 22′ 22.0″ + 2′ 58.2″ = N. 22° 25′ 20.2″. The difference is added because the north declination is increasing.

(d) 2. If the Solar Ephemeris gives the sun's declination at Greenwich APPARENT Noon, proceed as follows:

Sun's apparent declination at Greenwich APPARENT Noon on June 4, 1918 = N. 22° 22′ 21.4″. The difference in declination for 1 h = 18.03″. The equation of time is 2 m 1.51 s, and the difference in the equation of time for 1 h = 0.416 s. The place of observation is west of longitude 105° W. and 105th meridian time is used. 105° = 7 h. The standard time of the observation was 2 h 53 m P. M. = 2.883 h, which is 7 h + 2.883 h = 9.883 h after Greenwich Mean Noon. The equation of time at the instant of observation was 2 m 1.51 s —(0.416 s x 9.883)= 1 m 57.4 s = 0.033 h, which must be applied to standard time to obtain apparent time. The difference in declination at the instant of observation is 18.03″ x 9.883 + 0.033) = 178.8″ = 2′ 58.8″. The declination at the instant of observation is 22° 22′ 21.4″ + 2′ 58.8″ = N. 22° 25′ 20.2″. The difference is added because the north declination is increasing.

By previous observation, or from a map, the latitude of the place of observation has been determined = N. 39° 44′ 45″.

(e) 1. Computation by formula (2)

$$\cos \tfrac{1}{2} A_n = \sqrt{\frac{\sin S \sin (S - \text{codec})}{\sin \text{colat} \sin \text{coalt}}}$$

$$S = \tfrac{1}{2} (\text{codec} + \text{colat} + \text{coalt})$$

codec =	67° 34′ 40″
colat =	50° 15′ 15″
coalt =	40° 26′ 13″
2S =	158° 16′ 8″
S =	79° 8′ 4″

log sin S	= 9.9921434
log sin (S — codec)	= 9.3017612
colog sin colat	= 0.1141368
colog sin coalt	= 0.1880158
	2)9.5960572
log cos ½ A$_n$	= 9.7980286
½ A$_n$	= 51° 5′ 24″

A$_n$	=	102° 10′ 48″
horizontal angle	=	294° 54′ 30″
		397° 5′ 18″
		— 360°
Bearing	=	N. 37° 5′ 18″ W.

(e) 2. Computation by formula (12)

$$\cos A_s = \frac{+\sin dec}{\cos lat \cos alt} - \tan lat \tan alt$$

log sin dec	=	9.5814136
colog cos lat	=	0.1141368
colog cos alt	=	0.1880158
log 1st term	=	9.8835662
log tan lat	=	9.9198979
log tan alt	=	0.0694691
log 2nd term	=	9.9893670
2nd term	=	-0.975814
1st term	=	$+0.764832$
nat cos A_s	=	-0.210982
log cos A_s	=	9.3242454
A_s	=	77° 49′ 12″
horizontal angle	=	294° 54′ 30″

217° 5′ 18″
— 180°

Bearing   = N. 37° 5′ 18″ W.

Note: It is customary to make a series of five observations, compute the azimuth indicated by each, and take as the azimuth required the average of not less than three computations that check within one minute of arc. For this purpose formula (12) is the most convenient.

(e) 3. Computation by formula (15)

$$\cos A_n = \tan C_2 \tan lat$$
$$C_2 = \tfrac{1}{2} coalt - \tfrac{1}{2}(C_1 - C_2)$$
$$\tan \tfrac{1}{2}(C_1 - C_2) = \cot \tfrac{1}{2}(lat + dec) \tan \tfrac{1}{2}(lat - dec) \cot \tfrac{1}{2} coalt$$

alt =	49°	33′ 47″
coalt =	40°	26′ 13″
dec =	N. 22°	25′ 20″
lat =	N. 39°	44′ 45″

(lat + dec) = 62° 10′ 5″
(lat — dec) = 17° 19′ 25″
½ (lat + dec) = 31° 5′ 2.5″
½ (lat — dec) = 8° 39′ 42.5″
½ coalt = 20° 13′ 6.5″

log cot ½ (lat + dec)	=	0.2197847
log tan ½ (lat — dec)	=	9.1828120
log cot ½ coalt	=	0.4338048
log tan ½ (C₁ — C₂)	=	9.8364015

$$\frac{1}{2}(C_1 - C_2) \; = \; 34° \; 27' \; 17.6''$$
$$\frac{1}{2} \; \text{coalt} \; = \; 20° \; 13' \; 6.5''$$

$$C_2 \; = \; -14° \; 14' \; 11.1''$$
$$\log \tan C_2 \; = \; 9.4043466$$
$$\log \tan \text{lat} \; \pm \; 9.9198979$$

$$\log \cos A_n \; = \; 9.3242445$$

$$A_n \; = \; 102° \; 10' \; 48''$$
$$\text{horizontal angle} \; = \; 294° \; 54' \; 30''$$

$$397° \quad 5' \; 18''$$
$$-360°$$

$$\text{Bearing} = \quad \text{N.} \; 37° \quad 5' \; 18'' \; \text{W.}$$

2. By Equal A. M. and P. M. Altitudes of the Sun.

(a)   Set up transit over one end of line whose azimuth is desired. Sight along the line with the verniers at 0°. With the lower clamp tightened and the upper clamp loosened sight on the sun. If the upper and left-hand limbs are sighted in the forenoon, then sight on the upper and right-hand limbs in the afternoon. Use the same vertical angle in both observations, and record the horizontal angle and the time in each case. The mean of the two horizontal angles, corrected for the effect of change in declination, is the desired azimuth from the south.

(b) The angle between the meridian and the mean of the two horizontal angles is found by the formula:

$$\text{Correction} = \frac{\text{Half the change in declination between the two observations}}{\text{cos lat} \times \text{sin half the hour angle between the two observations}}$$

EXAMPLE:

$$\text{Latitude} = \text{N.} \; 39° \; 45' \; 36'' \qquad \text{Date} = \text{July 11, 1918.}$$

Observations:	A. M.	P. M.
Angle on desired course=	0°	0°
Vertical angle on upper limb=	63° 18'	63° 18'
Horizontal angle =	240° 3' (left)	352° 18' (right)
Time of observation=	10h 30m	1h 12m

Half the time between observations, or hour angle.
$$= 1\text{h} \; 21\text{m}$$
$$= 1.35\text{h}$$
$$= 20° \; 15'$$

Half the change in declination $= 19.37'' \times 1.35\text{h} = 26.15''$

$$\log 26.15'' \; = \; 1.4174717$$
$$\text{colog cos lat} \; = \; 0.1142261$$
$$\text{colog sin } 20° \; 15' \; = \; 0.4607770$$

$$\log \text{correction} \; = \; 1.9924748$$
$$\text{correction} \; = \; 98.282'' \; = \qquad 1' \; 38''$$
$$\text{mean horizontal angle} \; = \; 63° \; 49' \; 30''$$

$$\text{corrected angle} \; = \; 63° \; 47' \; 52''$$
$$\text{Bearing} \; = \; \text{S.} \; 63° \; 47' \; 52'' \; \text{W.}$$

Note: It is customary to take a series of observations in the fore-noon at suitable intervals, and corresponding observations in the after-noon, in order to check their accuracy and increase precision.

3. By Observing Polaris at Elongation.

(a) Set up transit over one end of the line whose azimuth is desired, about half an hour before elongation, and sight along the line with the verniers set at 0°. The standard time of elongation may be found by interpolation from a table of the time of elongation of Polaris. If such a table is not available, then the hour angle may be computed by the following formula:

$$\cos \text{ hour angle} = \frac{\tan \text{ latitude}}{\tan \text{ declination}}$$

This hour angle may be converted into sidereal time by the following formula:

Sidereal time = hour angle + right ascension.

This sidereal time may be converted into local mean time by the following formula:

Local mean time = sidereal time — mean sun's right ascension — in-crease in sun's right ascension.

This local mean time may be converted into standard time by ex-pressing the longitude between the local meridian and the standard merid-ian in units of time (15° = 1h), and adding this correction if the local meridian is west of the standard meridian, or subtracting the correction if the local meridian is east of the standard meridian.

The declination and right ascension of Polaris, and the mean sun's right ascension and the increase in sun's right ascension, may be found in the Nautical Almanac.

(b) Focus on the star and follow it with the vertical cross hair as it moves towards its greatest elongation. Near the elongation the star appears to move vertically. When the desired elongation is reached read the horizontal angle.

(c) From a table of azimuth of Polaris at elongation determine the azimuth corresponding to the latitude of the place of observation. If such a table is not available, then the azimuth may be computed by the fol-lowing formula:

$$\sin \text{ azimuth} = \frac{\sin \text{ polar distance}}{\cos \text{ latitude}}$$

$$\text{or } \sin \text{ azimuth} = \frac{\sin \text{ codeclination}}{\cos \text{ latitude}}$$

(d) Bearing = horizontal angle + azimuth at W. elongation.

or Bearing = horizontal angle — azimuth at E. elongation.

EXAMPLE:

(a) Transit is set up over point A, July 16, 1918, in latitude N. 39° 44' 45". From a table of elongations of Polaris the time of western elon-gation is found by computation to be 11h 59.3m P. M. Point B is sighted with the verniers set at 0°.

(b) With the lower clamp tightened and the upper clamp loosened the star is observed at western elongation, and the horizontal angle is 86° 47' 30".

(c) From a table of azimuth of Polaris at elongation the azimuth for latitude N. 39° 44′ 45″ is 1° 28′ 27″. Or the azimuth may be computed as follows:

Polar distance = 1° 8′ 2″.	log sin polar distance	= 8.2964195
Latitude = 39° 44′ 45″.	log cos latitude	= 9.8858632
	log sin azimuth	= 8.4105563
	azimuth	= 1° 28′ 27″

(d) Horizontal angle =   86° 47′ 30″
Azimuth at W. elongation =   1° 28′ 27″

Bearing of line A-B =   N. 88° 15′ 57″ W.

---

4. By Observing Polaris at Culmination.

(a) *C*ompute the exact standard time of culmination, and provide a watch reading correct standard time. Set up transit, before upper or lower culmination, over one end of the line whose azimuth is desired. Sight along the line with the verniers set at 0°. With the lower clamp tightened and the upper clamp loosened observe the star.

(b) Focus on the star and follow it with the vertical cross hair until an assistant reading the watch calls the time of culmination. The horizontal angle is the desired azimuth from the north.

EXAMPLE:

(a) Transit is set up before lower culmination on June 1, 1918. From a table of culmination of Polaris, the local mean time of lower culmination is 8h 51m 18s P. M. for longitude 90° W. For longitude 105° 32′ 33″ W. the local mean time would be (0.16m earlier for each 15°) 8h 51m 8s P. M. The longitude between the local meridian and the standard meridian (105°) is 32′ 33″, which expressed in units of time (15° = 1h) = 2m 10s. Standard time = 8h 51m 8s + 2m 10s = 8h 53m 18s.

(b) The horizontal angle at 8h 53m 18s P. M. is 88° 16′. Hence the desired bearing is N. 88° 16′ W.

---

5. By Observing Polaris at Any Hour Angle.

(a) Set up transit, at any time when Polaris is visible, over one end of the line whose azimuth is desired. Sight along the line with the verniers set at 0°. With the lower clamp tightened and the upper clamp loosened observe the star.

(b) Focus on the star and follow it with the intersection of the vertical and the middle horizontal cross hairs. Take a series of readings and record the time, horizontal angle, and vertical angle for each observation. Determine the index error of the transit if necessary. If the instrument is not in perfect adjustment, make the observations in pairs, with telescope direct and inverted, and average the two sets of angles, and determine the mean time.

(c) The azimuth may be computed by the following formulæ:

$$\sin \tfrac{1}{2} \text{ hour angle} = \sqrt{\frac{\sin \tfrac{1}{2}(\text{coalt} + \text{lat} - \text{dec})\sin \tfrac{1}{2}(\text{coalt} - \text{lat} + \text{dec})}{\cos \text{lat} \cos \text{dec}}}$$

$$\text{and } \tan \text{azimuth} = \frac{\sin \text{hour angle}}{\cos \text{lat} \tan \text{dec} - \sin \text{lat} \cos \text{hour angle}}$$

EXAMPLE:

Transit is set up at 8:30 P. M., June 27, 1918. Observations are made from 8:40 P. M. to 9:20 P. M. At 9:00 P. M. standard time in longitude 105° 32' 33" W., the observed horizontal angle was 88° 56', and the observed vertical angle was 38° 45'. The refraction correction is 1' 12". Hence the true altitude is 38° 45' — 1' 12" = 38° 43' 48".

Coaltitude = 90° — 38° 43' 48" = 51° 16' 12"

Latitude = N. 39° 44' 45"

Declination = 90° — 1° 8' 3" = 88° 51' 57"   (Refer to table of polar distances or declinations of Polaris)

Computation for hour angle:

½ (coalt + lat — dec) = 1° 4' 30". log sin = 8.2732604
½ (coalt — lat + dec) = 50° 11' 42". log sin = 9.8854899
                                  colog cos lat = 0.1141368
                                  colog cos dec = 1.7034741
                                  _____
                                  2) 9.9763612
                                  _____
          log sin ½ hour angle = 9.9881806

                   ½ hour angle = 76° 41' 36"
                      hour angle = 153° 23' 12"

Computation for azimuth:

          log sin hour angle = 9.6512462

log cos lat = 9.8858632
log tan dec = 1.7002091
_____
    1.5860723 = log 38.55426

log sin lat = 9.8057611
log cos hr = 9.9513618
_____
    9.7571229 = log   .57164
              log 37.98262   = 1.5795849
              log tan·azimuth   = 8.0717613
     azimuth east of meridian. = 0° 40' 33"
     horizontal angle to star   = 88° 56' 0"
     _____
     horizontal angle to pole = 88° 15' 27"
     Bearing of line   = N. 88° 15' 27" W.

Note: Culminations of Polaris for latitude, or elongations of Polaris for azimuth, may be observed without knowledge of the time if advantage is taken of the fact that Zeta Ursa Majoris (the star at the bend in the dipper handle), the north pole, Polaris, and Delta Cassiopeiæ (the star at the bottom of the first stroke of the W) are nearly in a straight line, with Polaris between the pole and Delta Cassiopeiæ. When this line is horizontal Polaris is at elongation, and when the line is vertical Polaris is at culmination, the elongation or culmination being in the direction towards Delta Cassiopeiæ.

CPSIA information can be obtained
at www.ICGtesting.com
Printed in the USA
LVHW021515261118
598291LV00012B/1085